Copyright © 2015 by Logic Version Tech Inc

All rights reserved.

Published by Logic Version Tech Inc

No part of this book may be reproduced or transmitted into any form or by means, electronic or mechanical, including photocopying, recording or by any information storage and retrieval system, without written permission from the publisher. For Information contact Logic Version Tech Inc,

Email: info@lvtinc.com

First Edition

ISBN 978-1-329-19526-4

Book design by Lavanya Rangaraman.

This edition first printing, June 2015
Printed in the U.S.A

Medicinal Plant – Vilva
A review on Aegle Marmelos

By Lavanya Rangaraman

This book is not intended as a substitute for the medical advice of physicians. The reader should regularly consult a physician in matters relating to his/her health and particularly with respect to any symptoms that may require diagnosis or medical attention. The information provided in this book is designed to provide helpful information on the subjects discussed. This book is not meant to be used, nor should it be used, to diagnose or treat any medical condition. For diagnosis or treatment of any medical problem, consult your own physician. The publisher and author are not responsible for any specific health or allergy needs that may require medical supervision and are not liable for any damages or negative consequences from any treatment, action, application or preparation, to any person reading or following the information in this book. References are provided for informational purposes only and do not constitute endorsement of any websites or other sources.

Thanks to my ancestors, my parents, my family and my teacher for guiding me in this endeavor.

Acknowledgements

With due respect, I dedicate this book to all the researchers, from my ancestors to the future discoverers around the world who have contributed to herbal medicine.

The contents of this book were educated to me through various people, sources and media who have thrown in their time and effort in bringing out the potential of Vilva to the world. As years passed the particulars of some sources were missed so as not to omit them, no particular reference has been mentioned in this book.

I salute all the practitioners, scientists and gardeners throughout the world for utilizing Vilva towards the well being of mankind.

Contents

1. Introduction — 11
2. Plants in Herbal medicine — 12
3. Vilva or Bael — 14
4. Uses — 15
5. Nutritional value — 16
6. Roots — 17
7. Barks — 18
8. Leaves — 19
9. Fruits — 21
10. Flowers — 23
11. Seeds — 24
12. Chemical compounds — 25
13. Medical conditions — 29
14. Cultivation — 31

Introduction

Traditional medicine (TM)

According to World Health Organization, Traditional medicine is termed as the oldest existing therapeutic systems used by humanity for health and well being.

Herbal medicine

A medical system which uses various parts of a plant and its extract to treat medical conditions is termed as Herbalism or Herbology.

History of Herbal medicine

Evidences on the use of herbal medicine can be traced back to 5000 years from the Sumerian culture.

The use of herbs back dates to Vedic period in India. The Vedas confirms the medical facts and the usage of the herbs in Rig, Yajur and Atharvana Veda. Charaka Sanmihta, the traditional book of Ayurvedic medicine provides the procedures and methodologies in administering herbs for treating medical conditions.

Various cultures around the world have been practicing herbal medicine for centuries. Some of the notable traditional systems around the world are

East Asia - Chinese, Japanese, Korean, Mongolian, and

Tibetan etc.
Southeast Asia - Indian, Sri Lankan, Thai and Vietnamese etc.
Mediterranean - Egyptian, Greek, Roman, Islamic and Unani.
Africa - Iboga, Nganga, African and Yoruba etc.
America – Aztec, Kallawaya and Brazilian etc.

Plants in Traditional Medicine

Plants form the basic level of the food web. They are the resources of Food, Fertilizer, Pharmaceutical and Paper industries. Various parts of the plants are used as ingredients in the traditional practices. Some of them are flowers, leaves, bark, seeds and roots. Herbs and spices were used in many cultures as preventive measures for diseases. They are taken in the form of teas, juices and capsules.

Some of the plants that are used in day today life are

- Ginger
- Lemon
- Cilantro/Coriander
- Turmeric
- Cinnamon
- Garlic and
- Mint etc.

Herbal Practitioner

People are using herbs and spices in their daily diet as defensive agents in combating common diseases like cold, fever and pain etc. The use of these spices is mostly through their family heritage rather than a formal education. So the herbs are not utilized to its full potential.

Practitioners often exercised well-built intellectual and historical ancestry towards traditional medicine. In ancient times this knowledge and wisdom was passed on to the successors from generation to generation. At one point in time when people stopped practicing due to economical constraints, they failed to pass on the knowledge to their hereditary. This is when the use of herbal medicines started to decrease.

India has a long heritage of using plants for medical purposes. Farmers primarily cultivate herbs and spices for their medical benefice. The population of herbal medicine is growing due to its low side effects, cost-effective and usefulness in the recent times.

Vilva or Bael

Introduction

Vilva is a native plant of the Asian sub continent. Vilva is mostly found in India, Sri Lanka, Pakistan, Thailand and various other countries in the continent. Vilva has been referred in India by different names like Bael, Bilva, Bela and Maredu etc.

Vilva is a subtropical plant and has a reputation for its ability to grow in places where other plants may not be grown successfully. In India they are widely cultivated in plains and mountain regions from Kashmir to Kanyakumari.

Vilva is a slow growing medium tree that bears fruits once a year during summer. It grows up to 12-15m tall with short trunk. This plant manages the soil with pH range from 5 to 8 and tolerates temperatures from -6°C to 48°C. It's a slow growing tree as it takes 11 months for the fruit to fully ripe.

Vilva belongs to the Rutaceae family (the family of citrus fruits) and botanically called as *Aegle Marmelos (L.) Correa*. It is the only plant that belongs to the genus Aegle.

Vilva is more famous for its medicinal and mythological uses than for edible consumption in India.

Medicinal Uses

Various parts of the tree is used in treating medical conditions related to heart, abdomen, digestion, cancer, fungal, and bacterial infections etc.

Mythological Uses:

Vilva is considered sacred in India and is cultivated throughout the country in the temple areas due to its mythological status. Vilva leaves are considered auspicious to Lord Shiva and are offered while praying. Lingam is showered with Vilva leaves and is believed that the devotee is relieved from the health ailments and thus the name *Shivadruma* has been given to this tree.

Edible Uses:

Various parts of this plant are consumed in different forms. The ripe fruits are used in preparing smoothies popularly known as "Sherbet" in India, the pulp as marmalade or jam, the leaves and shoots are eaten as a vegetable in Thailand and as seasoning in Indonesia.

Cosmetic Uses:

The dried peel of the fruit is used as lip balm. Due to its aromatic nature the extracts of the plants is used in perfumes. The diluted pulp of the fruit can be used as a skin cleanser and a skin toner.

Various studies have been made on the medicinal uses of the Vilva tree world wide and are found to be an underutilized plant with great potential in the pharmaceutical industry.

The Nutritional chart shows the values of the Vilva fruit (100gm).

Nutrition	Amount
Carbohydrate	31.8 gm
Fat	0.3 g
Proteins	1.8 gm

Vitamins	Amount
Vitamin A	55 mg
Vitamin B	Rich in Vitamin B1 and B2
Vitamin C	55 mg
Thiamine	0.13mg
Riboflavin	1.19 mg
Niacin	1.1 mg
Carotene	55 mg

Minerals	Amount
Calcium	85 mg
Potassium	600 mg
Fiber	2.9 gm
Water	61.5 g
Energy	137 k.cal

Various ethno medical literatures reveal the following uses of the plant in treating numerous diseases and conditions. Some of them are,

Roots

Roots of the Vilva tree are sweet and are used in treating various conditions stated below.

- The extract of Vilva root was studied for the inhibitory activity against *diarrhea* and was found effective in controlling the frequency of the stools.
- Vilva roots are one of the ingredients in the drug Dasamool (10 roots), an Ayurvedic drug used for treating *colitis, dysentery, diarrhea, hearing loss and fever.*
- The extract of the root has anti bacterial activity against *Cholera, Pneumonia, and E.coli and Typhoid* etc.
- *Boils, Sinusitis* and *food poisoning* can also be treated.
- The dried roots are used in treating *nervous system disorders, Edema, vomiting and rheumatism.*
- The bark of the roots can be used in treating *Insomnia, Hysteria, Seizures and Typhoid.*
- Also used as an *antidote* to snake venom in India.
- The studies on anti diabetic activity on the roots showed positive results. Also due to the hypoglycemic activity in the alcoholic extracts of the roots, this can be used in treating *diabetes*.
- The anti tumor potential of the roots can be used in treating *benign tumors* and *cancer cells*.
- Presence of antioxidant properties makes this plant a potential *antioxidant*.
- Can be used to treat inflammations and also as a *wound healer*.
- The extracts from the root has been used against *malaria* and *allergic rhinitis*.

- The root can be used in treating *anorexia, flatulence* and *colic*.

Barks

Bark of the Vilva tree is bluish–grey and soft with irregular furrows on the younger branches.

- The extract of bark was studied for the *anti diabetic* activity and also treating s*tomatitis*.
- Stem barks are used as a pain reducer and fever reducer due to its *antipyretic and analgesic* property.
- Used as an *antidote* and also in treating *diarrhea*.
- The extracts of the bark is used in treating *malaria*.
- The bark is also used as a *fish poison*.
- The immature bark of the tree can be used for treating *cancer*,

water retention and lowering the *body temperature*.
- Can also be used in treating *seizures* caused by *epilepsy*.
- The barks can also be used in treating *drug abuses*.
- This can also be used as *sedatives* and to treat *gingivitis*.
- Twigs are used to treat *gingivitis*.

Leaves

The leaves are alternate, ovate, trifoliate and aromatic. The leaves are borne singly or in 2's or 3's, are also composed of 3 to 5 oval, the terminal one with long pointed ones. New foliage is glossy and pinkish-maroon. Mature leaves emit an odor when bruised. Various studies proved the leaves possess enormous medicinal qualities and has a great potential for mankind.

- Traditional uses include treating *jaundice*, treatment of *wounds*, *leucorrhoea*, *conjunctivitis*, *styes* and *deafness*.
- The leaf extracts can be used in treating *diabetes*.
- The dried powder is used in treating *irritable bowel syndrome*.
- The oils extracted from the leaves can be used for treating *fungal infections*.
- The leaves when studied possessed antioxidant qualities and can be used as *antioxidants*.
- The poultice form of the leaves can be used in *ophthalmic* treatments and in treating *edema*.
- *Catarrhal fever* can be treated using the leaves which destroys phlegm and also in *dropsy*.

- The leaves form a good cardio tonic and used to treat *inflammations*.
- The juices can be used as a *laxative*.
- The effective uses for *ulcer* are proven. The leaf infusion can be used in treating *peptic ulcer*.
- Decoction can be used in treating *asthma* and can be used as an *expectorant*.
- Can also be used as *insecticide* against brown plant hopper.
- Vilva leaves are claimed to be used in treating *respiratory* and *cardiac* disorders.
- Cardiac protective property makes this an effective *cardiac depressant* and in treating *dyslipidemia*.

- Due to its anti-spermatogenic property, the leaves can be used in *birth control* drugs.
- Other uses include treating *intestinal parasites, antiseptic, anti-allergic* and *antibacterial*.

Warning: The leaves are said to cause abortion and is not recommended for pregnant woman.

Fruit

The fruits are yellowish green and have small dots on the outer surface. Fruits give freshness and energy. They come in 5 different shapes namely oval, flat, spherical, pear and oblong. The fruit is the more valuable part of the tree and a cooling drink popularly known as "Sherbet" is produced.

Unripe Fruit:
- The dried fruit pulp can be used in treating *constipation, amoebic dysentery* and *chronic diarrhea*.
- Vilva pulp extract can be used to treat *cancer*.
- Dried fruit powder can also be used in healing *burns* and *abrasions*.
- The unripe fruit is *astringent* and is used to cure *digestive ailments*.
- Appetite can be improved and so can be used as an *appetite stimulant*.
- The oil obtained by soaking the pulp with sesame oil is used to heal the *burning sensations* on the body.
- Due to the heat producing potency, it is used in treating *arthritis*.
- The decoction can be used along with other herbs in treating *hemorrhoids*.
- Can be used in *calico* and *silk fabric printing*.
- Due to its cardio protective property, extracts of the unripe fruit can also be used as *cardiac depressant*.
- The extracts can be used in treating *Ranikhet disease*.

Ripe Fruit:
- Vilva pulp extracts can be used as *laxatives*.
- The dried fruit powder can be used in treating *thyroid* disorders, *anemia, typhoid, coma* and *colitis*.
- The ripe fruit has more *antioxidant* compared to the unripe fruit.
- Dried fruit powder can also be used in treating *irritable bowel syndrome, fractures, swollen joints, bleeding sores* and *cramps*.
- The pulp is used as a *detergent* by substituting with soap.
- The mature fruits are made into jam, marmalade or syrup and

used in treating *diarrhea* and *dysentery*.
- Famous Vilva fruit *toffee* has been made from the fruit pulp.
- The ripe fruit is used in *indigestion (dyspepsia)*.
- Pulp is used in treating *scurvy* (anti-scorbutic) and also can be used as a *fever reducer*.
- Fruit extracts are used to treat *inflammations*.
- The pulp is also used to treat *leucoderma*, the extracts wards off *heat stroke* and other *heat afflictions*.
- In Burma they are used in making *paints*.
- Vilva fruit extract can be used in *lowering the level of cholesterol*.
- They can also be used to treat *viral and bacterial infections*.
- The beverage can be used in treating *ulcers*.
- The pulp extract can be used in treating the *intestinal parasites* and expel them from the body.
- They can also be used in treating *hypertension* and also as *heartbeat inhibitors*.
- Vilva fruit can also be used to suppress *muscle spasms*.
- Other uses include treating *bile disorders* etc.
- The in vitro activity was found to be comparable to that of ciprofloxacin. So it can also be used in *antibiotics*.

Flowers

- An infusion of the flower can be used as a *cooling drink*.
- They can be used as *astringent* and *antiseptic*.
- The diluted and distilled extract from the flowers can be used to treat *catarrh conjunctivitis*.

Seeds

There are numerous seeds embedded in the pulp of the fruit, oblong with cotton like hairs on their outer surface.

- The extracts from the seeds can be used as an *antibacterial*.
- The mucilaginous material around the seed can be used as an *adhesive*.
- Seed extracts can be used in treating *diabetes*.
- They can also be used in treating *gastric ulcers* and *malaria*.

Chemical constituents of Vilva

There are many chemicals present in various parts of the plant. Some of these compounds with their bioactivities are charted below.

Root and Bark contains skimmianine, fagarine, marmin, psoralen, xanthotoxin, scopoletin and tembamide.

Skimmianine	Sedative
	Hypothermic
	Anti diuretic
	Anticonvulsive
	Antipyretic
	Analgesic
	Anticancer
	Anti-methamphetamine
	Anti-malarial
	Hypothermic
Fagarine	Abortifacient
Marmin	Antiulcer
Psoralen	Antihelminthic
	Photo sensitizer
Xanthotoxin	Fish poison
Scopoletin	Antibacterial
Tembamide	Hypoglycemic

Flower and Seed contains luvangetin, palmitic, stearic, oleic, and linoleic acid.

Luvangetin	Gastro protective
Palmitic acid	Fatty acid synthesis

Stearic	Anti viral
	Anti inflammatory
Oleic acid	Adhesive
Linoleic Acid	Fatty acid synthesis

Leaves contains alkaloids, cardiac glycosides, terpenoids, saponin, tannins, flavonoids, steroids, anthocyanin, coumarins lignans, catechins, isocatechins, skimmianine, aegelin, lupeol, cineol, citral, citronellal, cuminaldehyde, eugenol, marmesinin, sterols and phlobatannins.

Skimmianine	See Root and Barks
Flavonoids	Antioxidant
	Anti-inflammatory
	Anti-tumor
	Antithrombotic
	Anti-allergic
	Antiviral
Coumarins	Anti fungal
	Anti tumor
Anthocyanin	Antioxidant
	Cardio protective
Saponin	Antioxidant
	Immunity stimulant
	Antitumor
Lupeol	Antiprotozoal
	Anti-inflammatory
	Antitumor
	Nutraceutical/Chemo protective
	Cardio protective
	Hepatoprotective

Sterols (Beta-Sitoserol)	Antimicrobial Anti allergic Anti-aging Antiserum/Antidote Antiurolithiatic Antifertility Gastro protective Anticonvulsive Antitumor Antioxidant Atherosclerosis Prostate enlargement
Eugenol	Antiulcer Antibacterial Antioxidant Hepatoprotective Anti-infective
Lignan	Antioxidant
Marmesinin	Antioxidant Cardio protective
Aegeline	Antihyperlipidemic Antioxidant Hypoglycemic Cardio active
Terpenoids	Antibacterial Anti fungal
Cineol	Anti infective Antiulcer
Citral	Anti allergic Antiseptic
Citronellal	Antiseptic

Fruit and fruit pulp contains flavonoids, steroids, terpenoids, phenolic compounds, lignin, fat, oil, insulin, proteins, carbohydrates, alkaloids, cardiac glycosides, minerals, vitamins, marmelosin, luvangetin, tannin, psoralen, aurapten and marmelide.

Marmelosin/ Marmelide	Antibacterial Antihelminthic
Flavonoids/ Terpenoids	See Leaves
Luvangetin	See Seeds
Psoralen	See Roots and Barks
Tannin	Anti diarrhea Astringent
Aurapten	Heartbeat inhibitor

Various alkaloids like aegelenine, marmeline, fragine and dictamine, coumarins like imperatorin, marmin, alloimperatorin, xanthotoxol, methyl ether, umbelliferone, marmenol, scoparone, and polysaccharides like galactose, arabinose, uronic acid and L-rhamanose, caratenoids were found from various parts of the plant.

Ascorbic acid, sitoserol, crude fibers, α-amyrin, crude proteins are also found in minor quantities.

Medical conditions and diseases treated but not limited to are

Abdominal pain	Abrasions	Allergic rhinitis
Anemia	Antidote	Anxiety
Appetite Stimulant	Arthritis	Asthma
Athlete's foot	Bacterial infections	Bile disorders
Birth control	Bleeding gums	Body ache
Boils	Bronchitis	Burning sensation
Burns	Cancer	Cardiovascular disorders
Catarrhal fever	Cholera	Cholesterol
Colic	Coma	Common cold
Conjunctivitis	Constipation	Convulsion
Cramps	Diabetes	Diarrhea
Dropsy	Drug abuse	Dysentery
Dyspepsia	E.coli infections	Edema
Epilepsy	Fever	Fractures
Gastric trouble	Gingivitis	Gonorrhea
Hearing loss	Heart attack (myocardial infarction)	Heat stroke
Hemorrhoids	Hyperlipidemia	Hypertension
Hyperthyroidism	Hypochondiasis	Indigestion
Inflammation	Insecticide	Intestinal infections
Jaundice	Jock Itch	Laxative
Leucoderma	Liver ailments	Lowering body temperature
Malaria	Melancholia	Nausea
Ophthalmia	Pain	Palpitation
Peptic Ulcer	Pneumonia	Ranikhet disease

Respiratory disorders	Ringworm	Scurvy
Seizures	Skin diseases/ infections	Stomachache
Styes	Swellings	Swollen joints
Tinea infections	Tumors	Typhoid
Ulcer	Urinary tract infections	Viral infections
Water retention	Worms and Parasites infections	Wound infections

Cultivation

Soil and Climate

Good sandy soil will be best suited for this plant but has also grown successfully in swampy, alkaline and stony soils. The pH range varies from 5 to 8.

Sunny and warm humid climate with 20°F to 120°F are suitable climatic conditions for cultivation.

Sowing - Vilva can be grown through seeds. It is a slow growing tree and takes quite sometime for the seedlings to develop. Planting in the nursery will be a best option for the first year. Then transplanted into the field would be beneficial so that the nematodes would not be destroyed. First year requires more attention with regards to manure and weeding.

The choice of fertilizers should be given maximum consideration as this is a medicinal plant. Organic manures and natural manures are some choices to be considered.

The tree yields around 200 fruits average and may go up to 800 in a season. This tree also grows as a container plant for home gardening or as a greenhouse project.

Based on the ethno medical evidences, cultivating Vilva on a large scale will be very helpful to mankind. In my opinion Vilva has a great potential for the pharmaceutical industry. Extensive research carried out on this plant may discover more uses to human well being than indicated in this book.

I invite you all to come forward to plant a Vilva tree and enjoy the medicinal values of this *Plant of Panacea.*

Let us make a difference not only in the lives of present but also in the lives of the future generation.

www.ingramcontent.com/pod-product-compliance
Lightning Source LLC
Chambersburg PA
CBHW072310170526
45158CB00003BA/1259